HOW TO WRITE A SCIENTIFIC PAPER

HOW TO WRITE A SCIENTIFIC PAPER

AN ACADEMIC SELF-HELP GUIDE FOR PHD STUDENTS

JARI SARAMÄKI

Copyright © 2018 by Jari Saramäki

All rights reserved.

No part of this book may be reproduced in any form or by any electronic or mechanical means, including information storage and retrieval systems, without written permission from the author, except for the use of brief quotations in a book review.

For my students

CONTENTS

Who Should Read This Book And Why?	1
Why Does Writing Feel So Hard?	4

PART I
STORY

1. How To Choose The Key Point Of Your Paper	11
2. How To Choose The Supporting Results	16
3. How To Write The Abstract	20
4. How To Choose The Title	25

PART II
OUTLINE

5. The Power Of Outlining	31
6. How To Write The Introduction, Part I: Structure	35
7. How To Write The Introduction, Part II: A Four-Paragraph Template	38
8. How To Write The Introduction, Part III: The Lede	43
9. How To Write The Materials And Methods	47
10. How To Write The Results, Part I: Figures	52
11. How To Write The Results, Part II: Text	58
12. How To Write The Discussion	62

PART III
WORDS

13. How Does Your Reader Read?	69
14. How To Write Your First Draft	74
15. How To Edit Your First Draft	78
16. Tips For Revising Content And Structure	82
17. Tips For Editing Sentences	86

PART IV
IT'S NOT OVER YET

18. How To Write The Cover Letter	93

19. How To Deal With Reviews 97

Summary 104

Afterword 107
About the Author 111

WHO SHOULD READ THIS BOOK AND WHY?

This book has been written for the PhD student who is aiming to write a journal article on her research results, but any scientist who has ever found writing difficult might find something of use here.

Knowing how to write papers that other practitioners want to read and cite is an essential skill for any scientist, but it is rarely taught to students of science and technology. Writing is, however, an important part of our jobs. We are writers, whether or not we want to be, and the quality of our writing can help or hinder our careers. But somehow, we are supposed to pick up the necessary writing skills on our own, almost as if by osmosis. We are supposed to simply absorb them from somewhere.

This book was born from my attempts to fill this gap. It started with a series of blog posts at http://jarisaramaki.fi that became popular. These posts have been expanded upon and rewritten as the chapters of this book.

I chose to focus on PhD students and the specific task of writing a journal paper because this is a common struggle. PhD students are expected to quickly figure out both how to do research and how to write it up, and this can be rather stressful. Additionally, they rarely get enough help, in particular when it comes to writing. We professors tend to be too busy, and we often find it difficult to help our students because we struggle with writing too. Even those of us who are experienced writers may have no system for writing—and when there is no system, it is difficult to teach others what it is that we do and how do we do it.

After realising this, I decided to come up with a set of guidelines—a system for writing that I could follow with my students. The outcome is this book. What you'll learn here is a top-down approach that makes writing less taxing because it forces you to focus on the right things at the right time, one thing at a time.

Of course, to write a good scientific paper, you have to do some good science first. This book only covers the former. You cannot and should not compensate for bad science with good writing. But if you have done solid science, it deserves to be heard about, and it helps if you know how to communicate your results clearly and in a compelling way. It also helps if you can write quickly, without spending too much time being stuck or in fear of the blank page.

I have written this book mainly for students of science and technology (computer science, physics, life sciences, and so on) where the output of research typically comes in the form of plots, tables, and figures that lead to conclusions about how the world works. I wrote with a reader in mind who has already obtained research results that are worth publishing and who now wants to transform them into a well-written paper, a paper that has a streamlined story that is communicated through both text and figures. If needs be, this story can be condensed into the letter format of the PNAS, Nature, or Science variety. But getting your results published in the top-tier journals is not the main focus of this book: writing well for any journal is.

If you write well, you do the scientific community a service. Students, professors, journal referees, and other readers of papers are too often the victims of bad, convoluted writing that steals away their most valuable resource —their time. Just consider how many unnecessary hours you have spent trying to tease out the hidden meaning from research papers that carefully and jealously guard their secrets. We all have to deal with limited time: there is never enough of it, and wasted hours can never be claimed back. Save your readers' time by making your papers easy to read.

Finally, this book is not about language or grammar. If you wish to expand your vocabulary or improve your knowledge of grammar, please look elsewhere: there are many excellent resources available online and in the nearest bookshop (if there still is one). Rather, this book is about clarity, structure, excitement, and story. It is also a book about thinking because to write clearly is to think clearly.

WHY DOES WRITING FEEL SO HARD?

"Being a writer is a very peculiar sort of a job: it's always you versus a blank sheet of paper (or a blank screen) and quite often the blank piece of paper wins." — Neil Gaiman

Every scientist has struggled with writing at some point. Most of us struggle with every paper we write. Science is hard, and so is writing. Together they are harder. If you are a PhD student, you can add in a lack of experience as a researcher and as a writer. And when you combine all of that with the usual time pressure, it is no wonder that the blank document in front of you looks like the north face of Mount Everest. It looks impossible. We have all been there, staring at that wall.

While no mountaineer would risk climbing Everest without a route plan, an inexperienced writer tends to neglect the importance of planning. Having no plan, she tries to do everything at once. She opens a blank document in her editor. She stares at the document—still blank—and tries to think of the first word of the first sentence of the first paragraph. But because she doesn't yet know what story the first paragraph (or the whole paper) should tell, she tries to figure that out at the same time as choosing the first word. She doesn't necessarily even realise that this is happening. The first sentence just feels excruciatingly difficult to write. It feels like a looming vertical cliff covered in ice, too steep to climb. And that's just the first sentence.

At the same time, our writer is still—perhaps unbeknownst to herself—in the process of understanding her results and what to make of them. She might even be uncertain of the point that her paper should make. Everything is undecided. Everything is hanging in the air. The cliff is slippery. It is hard to get a grip on anything.

Under these circumstances, it is not surprising that writing feels impossible. It *is* almost impossible if you don't have a plan for how to proceed. No one can solve all these problems at the same time, from choosing the right words to figuring out what it is that you want to tell the world. Problems are best solved one at a time.

Writing becomes much easier if you separate the process of thinking from the process of writing. To write clearly is to think clearly, and for scientific papers thinking should precede writing. Writing becomes much less of a struggle if you first think through the right things in the right order.

Think of, say, a software project. At the beginning, the software architects and developers consider the big picture: what should the software do? What functions and classes are needed for it to work? No developer would ever begin by writing code for the internal bits and pieces of these functions and classes without knowing how they interface with the rest of the program. It is meaningless to write code for functions whose purpose and specifications are unclear.

Similarly, a successful paper-writing project should never begin at the level of words and sentences. Rather, it should begin at the level of ideas and structure. It is much more efficient to first consider what the point of the paper is and how to best communicate the results in terms of structure and storyline. Details such as sentences should be left for the final stages of writing.

Another way of looking at the problem of writing is linearity versus modularity. The fear of the blank page arises out of linearity: the feeling that the only way to fill the page is from left to right, starting with the first word and proceeding towards the last, sentence by sentence, word by word. This is not so. Whereas reading is usually linear, writing does not have to be. The process of writing should be modular: first, sculpt your raw materials into rough blocks that together form your story, and then start working on the blocks, filling in more and more details until entire sentences begin to appear towards the end of this process.

In this book, you will learn an approach wherein the writing process is divided into a series of hierarchical tasks. This makes it easier and less painful to get from a pile of results to a polished research paper.

This top-down approach begins by identifying the key point of the paper and then involves structuring the material that supports this point into a storyline. That's right: scientific papers *are* stories. They are not just containers of information! This storyline is then condensed into the abstract of the paper—my advice is to always write the abstract first, not last. This is unconventional, but it works. Writing the abstract first serves as an acid test: if you cannot do it, your storyline is not ready. But if you can, congratulations, because now you know the point of your paper and have a story to tell.

After finishing the first draft of the abstract, there are many steps to be taken before writing complete sentences again. These include planning the order of presentation and figures and mapping the story into paragraphs so that the topic and point of each paragraph is decided in advance. Then, the paragraph contents are expanded into rough sketches, and finally these sketches are transformed into whole sentences. At this point, there is no fear of the blank page because there are no blank pages. For each section and for each paragraph, there is a map and a route plan, and the only decision that you need to make is how to best transform that plan into a series of words.

There is an additional bonus for writers who have a good plan: it becomes possible to make progress whenever there is a gap in the schedule. When writing without a plan, it is very difficult to come up with much useful text if you unexpectedly get 15 minutes of spare time because your commuter train is delayed. If you have no plan, it takes all of those 15 minutes just to remember where the paper was supposed to be going. But when writing with a plan, you can just have a look at your notes and immediately start turning those notes into sentences. This way, you become much more productive.

When you have transformed your outline into text, the outcome of this top-down approach is a finished draft of the manuscript that already resembles the polished end product. But it shouldn't be too polished. It is usually much faster to write a quick-and-dirty first draft and then edit it several times than it is to attempt flawless sentences from the outset. Editing is much faster than writing and—at least for me—much less painful.

So, let's get started! I hope you now have a number of results at hand and at least a vague idea of their meaning. It is time to figure out together what it is that you want to tell the world!

PART I
STORY

1

HOW TO CHOOSE THE KEY POINT OF YOUR PAPER

"Most of the fundamental ideas of science are essentially simple, and may, as a rule, be expressed in a language comprehensible to everyone." — Albert Einstein

Let's begin our top-down approach to writing your paper at the very top. The first choice that you have to make is also the most important one: what is your paper about? What is its key point? What is its conclusion? Your answers lay the foundation for the rest of the paper.

Well-written papers are often about a single thing. Their key points are concise, and this makes them easier to read and understand. So when thinking about the point that you want to make, try to state it in one or two sentences. If you think that a few sentences is too little, consider the following: the Earth rotates around the Sun and not vice versa. Space-time is curved by mass. The salt of deoxyribose nucleic acid has a structure with two helical chains, suggesting a possible copying mechanism for genetic material. And so on.

The above sentences illustrate that there is often a lot of depth behind results that do not require too many words to describe. Choosing a key point that can be compressed into a few sentences does not mean that your paper has to be simplistic or overly commercial.

If you can condense your key point into a package that can be easily communicated, it is likely that it will eventually reach its intended target—the reader. This is not limited to primary transmission, whereby a potentially interested reader comes across your abstract and decides to read on. Secondary transmission is important too: turning the interested reader into someone who is excited enough to share your paper with her colleagues or to tweet about it. In both cases, it helps if your message is compact and focused. Communication is difficult and communication channels are noisy, so make it easy for your message to reach its audience in one piece.

Making a clear point is important, but what should this point be? If you have a notebook full of results, how do you select one to write about? This choice can be surprisingly difficult. Much of research is exploration, stumbling along a path that is hidden from sight. While this exploration can reveal many things, their meaning or importance is not always obvious. Often, the reality is that you have worked hard and produced some solid results, but nothing immediately stands out.

Results that look big and important at first sight are rare. Sometimes, however, the importance of a result only becomes apparent a long time after it has been published. It also helps to remember that all major breakthroughs build on smaller breakthroughs; they cannot and do not happen in isolation. The final piece of a puzzle can only be placed when the other pieces are already there, and some papers have to describe those other pieces. As such, not every paper can be amazingly important, and not every paper can be about the final piece. However, every paper can be focused, easy to understand, and come with a clear message.

If you have a number of plots, tables, and results at hand and are unsure as to which one to write about, try going through the results one by one. Imagine how you would describe a result to a colleague. How would it help others working in your field? Does it solve a problem that others are interested in? Does it open doors to new problems? Do you understand the result, or does it make you go "that's odd" (which is often a good sign)?

Try to explain the significance of each candidate result with full sentences, either by playing out imaginary conversations in your head or by talking to your colleagues or your supervisor. If you cannot find anything to say about a result, you might want to put it aside and move on to the next.

In my research group, we often do this exercise jointly with everyone involved in the paper-to-be. We reserve a few hours of time and go through the available results and plots on a meeting-room video screen or on the whiteboard. We discuss the meaning, interpretation, and importance of each result and consider whether it could lead to some new research questions. On many occasions, not understanding a result has led us to ask the really interesting questions. Therefore, it pays to be honest: if you don't get it, say so. If there is something odd in a plot, call it out. If you are not convinced by the explanations of others, say it. You might have spotted something that no one else gets either, and that could lead somewhere important.

But what makes a result great and exciting? While there are no general rules applicable to all fields of science, here are some aspects worth considering.

First, as discussed above, a good result is often one that is compact and focused enough to be compressed into one or two sentences.

Second, if a result feels surprising and unexpected when you see it for the first time, it deserves a closer look. If a result makes you go "that's funny", it probably contains a story that is worth telling. An unexpected result can even be turned into an exciting story by making a point out of not understanding it—"we have observed an unexpected something that we cannot explain". If the result is surprising enough, others will attempt to explain it for you.

Third, if your result solves a known, open problem, or if it can be used as a building block for a future solution, it is valuable to others in your field.

Fourth, if your result opens new problems, it is even more valuable—your fellow scientists will be happy to attack those problems and cite you while they are at it. Being a scientist is not about answers but about questions. A scientist without questions is an unhappy scientist—give others good questions and they will love you for it!

Fifth and finally, remember that beauty and elegance matter. Aesthetics are important, especially in the mathematical sciences—mathematical elegance is one of those things that you just know when you see it.

If no clear winners emerge after this exercise, then there are three alternatives.

First, should there be strong enough candidates among your results, you can always just pick the most promising one as the central result of your paper and see what happens if you go ahead. If after a while you find yourself at a dead end, then by all means do reconsider.

The second alternative is that you are simply not ready to write the paper yet, and you need to do more research to discover something worthwhile. This is perfectly OK, and it happens often—it is part of the process. In such cases, you just have to do a bit more work before you can return to writing.

The third, less likely option is that you have the materials for an exploratory paper in your hands—a paper that does not have a central result. An exploratory paper can be compared to a map of unknown terrain: there are lots of features that may be interesting, but no feature stands out. There may be hidden treasure, but finding it would require deeper exploration of some part of the map. It is not easy to write a good exploratory paper, but it can be done if necessary, especially as a follow-up to a more story-oriented paper.

If you are now lucky enough to have an exciting result that you think the world should know about, read on. In the next chapters, we will talk about how to put your result on a pedestal.

2
HOW TO CHOOSE THE SUPPORTING RESULTS

"Truth is ever to be found in simplicity, and not in the multiplicity and confusion of things." — Isaac Newton

Once you have decided on the key point of your paper, the next step is to choose what else goes in. This choice should be made with care. Now that you have a point to make, everything else should support this point. The best papers are often minimalistic: they drive their point home with essential ingredients only. Papers that contain lots of unrelated results are difficult to comprehend because they leave the reader wondering where to focus her attention. Clutter reduces clarity. Keep the material that makes your paper better and discard everything else.

The process of going through your results and deciding what to keep resembles the editing of a Hollywood movie. After the movie has been shot, the director and the editor start working with an abundance of raw materials that are to be sculpted into the final product—the theatrical cut. The goal is to assemble the film from the shots and scenes that best support the storyline while cutting out non-essential footage.

Your paper is your theatrical cut. Only use what it needs, and leave out the rest.

It may feel painful to discard some of your results—you spent a WEEK making that plot! But, believe me, it is for the best. If you want your paper to make an impact, others will have to read it and understand it first, and they will be greatly hindered if there is too much irrelevant information to absorb.

Perhaps it is because of the pain of discarding perfectly-good-yet-unimportant results that most journals nowadays allow for an extended director's cut in the shape of a Supplementary Information document with an unlimited page count. Authors can dump all those raw materials that didn't make the theatrical release into the supplement where they can be safely forgotten by the rest of the world (but at least the week spent making that plot feels like it meant something).

Let us see how far we can push the film industry analogy. A typical film script begins with the setup phase where the characters and the setting are introduced. Then it proceeds into the confrontation where the characters are put in trouble, and finally there is the resolution of the story (an epic fight in space followed by an exploding Death Star or something). This may be followed by a brief epilogue (with or without frolicking ewoks).

If we divide our results into these four plot phases, the setup category contains the results and figures that are required for the reader to make sense of your context, your experiment, and/or your data and its credibility (basic statistics and so on). It also contains schematic diagrams that visually explain the concepts your paper works with—always include a schematic diagram or two.

The confrontation category contains results that take the reader to the main result. These results bring the story closer to your final revelation; they build up excitement and lead the storyline toward its climax. You can build tension, for example, by showing results that are surprising and then revealing their explanation as the resolution of your storyline. You can also build up excitement by presenting competing hypotheses or models and asking which of them match your data.

The resolution category should contain only your key result, illustrated by no more than one or two plots or figures.

The last category of results—the epilogue—shows what follows after the resolution. The epilogue plot phase of a scientific paper is more important than the last couple of minutes of a blockbuster film. These results are presented after the main result has been introduced and serve the purpose of highlighting its significance. One key technique is to think of some application or consequence of the main result and to illustrate this with a figure or two.

If you look at some research papers published in the glossy magazines (Nature, Science, and so forth), you'll see that many authors apply this technique. Out of the four or so figures in those letter-format papers, the first is about setup and confrontation, the second is the key result (the resolution), and the rest are there to show why the key result matters or what it means (epilogue). In the kinds of journals where us mere mortals are published, there may be more figures per category, but the important thing is to set clear roles for your results and figures and use them accordingly when telling your story.

Armed with the above plot structure and the four categories, it should be easier for you to choose the results to include in your paper. Does a result play a clear role in one of the four plot phases—setup, confrontation, resolution, epilogue? Or, looking at it the other way around, what results should you include in, for example, the confrontation part to make it as strong as possible?

To summarise, at the end of this step you should have your key result at hand, together with three sets of auxiliary results: one that is essential for context (setup), one that builds up excitement and leads to the main result (confrontation), and one that highlights the importance of that key result (epilogue). These results, together with their order of presentation, determine much of your storyline.

Next, we will develop a super-compressed version of that storyline: the abstract of your paper.

3
HOW TO WRITE THE ABSTRACT

"Clear thinking becomes clear writing: one can't exist without the other." — William Zinsser, On Writing Well

We will now take your key result and your auxiliary results and develop the storyline further. The best way to do this is to take the bull by the horns and write the abstract of the paper.

It takes some courage to write the abstract this early, before any other part of the paper has been written, but it is worth it: the earlier you think through your storyline and put it into words, the better. The abstract is the storyline of the paper in miniature form. It determines the rest. Once you have composed your abstract, you have decided what story you want to tell. This makes the paper much easier to write and results in a more focused outcome; you can think of the rest of the paper as an extended version of the abstract.

The time you spend writing the abstract will pay off later. I've mulled over abstracts for several days, writing and rewriting. As well as determining your story, the abstract is critical for making your paper catchy. It is also the first thing that a potential reader sees.

If the abstract is badly written, unfocused, or missing essential elements, the reader cannot know why you have done the things that you have done or what your results mean. A messy abstract will not compel the reader to read the rest of the paper. A fellow scientist who randomly comes across your abstract will quickly move on if she cannot see what your point is or why your point is important.

A badly written, confused abstract is also an easy way of getting rejected quickly and harshly. The abstract makes the initial impression. It is the first part of your paper that the journal editor and the referees encounter—and it might be the last.

One common mistake is to view the abstract as a linear string of equally important pieces of information, whose only aim is to let the reader know what the author has done. An abstract written this way reads like a boring list of results: "We did X, and the result was Y. Then we did Z and..." Two important things are missing from such an abstract: context and excitement! Another very common mistake, especially among students writing their first papers, is to end the abstract abruptly after the results have been described. The reader is then left alone to figure out why these results matter.

So how is a great abstract structured? Is there a winning formula?

Yes, there is. By now you ought to have guessed already: great abstracts follow the tried-and-tested movie script structure of setup, confrontation, resolution, and epilogue. And I didn't make this structure up myself, as you will soon see. In fact, this exact formula is demanded by some of the top journals, because it works.

The task of the first few sentences of the abstract is to provide context and excitement. You can only follow the story if you understand the setting and know the characters (context), and you will only feel invested in the story if you care about the characters (excitement). The same applies to any research paper and its abstract: your reader has to understand the context and care enough about your research problem to read on and find out how the problem was solved.

After the setup phase, the script continues to the confrontation phase where there is trouble lurking, and where we encounter an issue that the characters have to solve. In the abstract, this section is the concise and focused description of the specific research problem that you have solved. It is immediately followed by the resolution, the high point of excitement in the story: your key result and the main conclusion of the paper.

The rest of the abstract should map to a (prolonged) epilogue of a film script. The last few sentences should illustrate how your results have changed the world, first within your field and then more broadly.

An abstract that follows this storyline resembles an hourglass: it starts by broadly introducing the setting in the setup phase. Then it narrows down to a specific research problem (confrontation) and its solution (resolution). Then the hourglass widens again as the abstract returns to the broader picture (epilogue).

It is not a coincidence that this is exactly how every Nature abstract reads.

The script that every Nature abstract has to follow, sentence by sentence, begins with a few lines on the general context and on the broader topic. These sentences should be comprehensible to "a scientist in any discipline", and they should describe the general interest in the topic. Then, the abstract narrows down to more specific context (again a few sentences) that can be slightly more exclusive and aimed at "scientists in related disciplines". This ends the setup phase. Next, the abstract funnels to its narrowest point, the confrontation: the exact research question that the paper addresses, stated in a single sentence. Then comes the resolution: the key result, again described in a single sentence. Finally, the abstract's scope becomes broader again as it enters the epilogue phase, first addressing the implications of the result to the paper's field of science and then discussing the general impact beyond that particular field, both in one or two sentences.

Even if you are not writing a Nature paper (and you probably aren't), the above provides a good recipe for an abstract. I write every single abstract I produce this way. If you follow the hourglass structure, your abstract will be focused and contain all the essential elements of a good scientific storyline. The broad beginning and end of the abstract force you to think about why you do what you do, which is never a bad thing, and having to concisely express the research problem and its solution makes you distill what you have done down to its essence.

Depending on your field and the chosen journal, the breadth of the top and the bottom of the hourglass may need to be adjusted. Rather than seeking to solve mankind's most pressing problems, the context of your results may simply be your particular field of science, or its subfield. For a specialist journal, you don't need to begin your abstract with a sentence on the importance of your field—the readers already know it. But it still it pays to consider the broadest context you can honestly think of. Don't exaggerate, but try to take a broader perspective. Why does your research question matter? Why are your results important? The answers to these questions should become the first and last sentences of your abstract.

4

HOW TO CHOOSE THE TITLE

"Any word you have to hunt for in a thesaurus is the wrong word. There are no exceptions to this rule." — Stephen King

After you have written your abstract, the next task is to consider the title of your paper. If the abstract is a compressed version of your storyline, the title of your paper is even more so. Titles are hard—it is often surprisingly difficult to come up with a short, informative, and catchy title. For me, this has at times felt like the hardest part of writing a paper.

The title of the paper serves a dual purpose: it delivers information by telling readers what your paper is about, and it serves as a marketing tool that makes others want to read your paper. Unfortunately, unlike the abstract, there is no general-purpose formula to follow when thinking of a title. There are, however, some points that you should consider.

The title has to be in perfect sync with the abstract—they have to tell the same story. Make sure that your title and abstract use the same words and concepts. Also make sure that everything that is mentioned in the title is discussed in the abstract.

Use words that everyone in your target audience can understand. Avoid subfield-specific jargon. Simply does it! The paper's title should only contain concepts that can be understood on their own, without any explanation. While there is some room in the abstract for explaining one or two important concepts in brief, there is no such luxury in the title: the reader should already be familiar with every word used in it.

The title should be focused and clear. If it is possible to give away the main result in the title, do so. Avoid vague titles, such as *"Investigating Problem X with Method Y"*. Instead, go for something more concrete: *"Investigating Problem X with Method Y Reveals Z."*

A small request: please never, ever use a title of the *"Towards Understanding Problem X"* variety. Just don't do it. Pretty please. If your research is worth publishing, you have arrived somewhere. Just be confident and tell the reader where this is, instead of telling them where you would rather have gone! It is OK to say something about the bigger picture in the title, as long as your key point plays a leading role. But to keep your title concise, it may be better to describe long-term goals elsewhere in the paper.

It helps if the title is catchy as well as informative. But do not exaggerate—consider how your title will look 10 years from now. Will it stand the test of time? If the title is too gimmicky or contains a joke that becomes stale after you've heard it a few times, it won't. You should also avoid jargon and buzzwords that may go out of fashion before the paper gets published.

Consider search engines and online search. Your paper needs to be found if it is to be read, so the title should contain the right keywords or search terms. As a network scientist, I almost always include the word "network" in my paper titles, even if this makes the title longer or if other network scientists would understand the title perfectly well without networks being explicitly mentioned. Without the word "network", they would not necessarily find my paper when they hunt online for new reading material.

Keep your title short. Research has shown that shorter titles attract more citations—see Letchford *et al.*, R. Soc. Open Sci. 2(8):150266 (2015). This should not come as a big surprise: long and cluttered titles are not as contagious as simple, focused ones. If the title is convoluted and hard to grasp, then the paper probably is too.

Sometimes there are field-specific conventions that you should be familiar with. In some biomedical fields, for example, the paper's title often expresses just the key result —*"Transcription Factor X is Involved in Process Y"*—and the titles can be fairly long. In some areas of physics and computer science, shorter and less informative titles are the norm. Have a look at other papers in your field, and try to imitate their best titles.

If you get stuck at this point and find it hard to decide on the title, it might be easier to initially lower your bar a bit. Just come up with some candidate titles that do not have to be perfect. Then ask your colleagues—your fellow PhD students, your supervisor, anyone—to have a look at the list and to pick the most promising candidates for refinement and final polishing.

PART II
OUTLINE

5

THE POWER OF OUTLINING

"Perdition awaits at the end of a road constructed entirely from good intentions, the devil emerges from the details and hell abides in the small print." — Iain M. Banks, Transition

Congratulations! If you have made it this far, you already have a vision of the story that you want to tell with your paper, and the essence of this vision is encapsulated in its abstract and title. You have your ingredients and raw materials at hand; you know your key result and your main conclusion. The rest of your results are organised into categories according to their role in the story: setup, confrontation, resolution, and epilogue. The basic elements are all there. And if everything is not yet crystal clear, please don't worry; your plan doesn't have to be perfect, as long as there is a plan.

The next step is to take what you have and start fleshing out the storyline by adding details. Because we follow a top-down approach, we will add these details layer by layer. The first layer is the outline of the paper that provides a map, a blueprint, and a skeleton for more layers to come.

The first task when outlining is to choose how to divide your paper into sections. Depending on your target journal, you may need to follow strict guidelines, such as the commonly used Introduction–Methods–Results–Discussion structure, or come up with a structure of your own. It pays to have some idea of what goes where, even for short letter-format papers that may or may not have section headings. Usually, this is not difficult: all papers begin with an introduction and end with a discussion, even if those elements span just a few paragraphs. The results are sandwiched somewhere in between. Methods may be explained before the results or after the discussion as an appendix of sorts (like in Nature and other glossy magazines).

Choosing the order of presentation within each section can feel more difficult, especially for the beginner. Luckily, there are some standard templates and structures you can follow that just work. In the coming chapters, I will discuss each of the usual sections in detail, from Introduction to Methods and from Results to Discussion, providing outlines for typical structures or ways of coming up with your own structure. For the Introduction, I will suggest a paragraph-level template that you can follow. For Results, a solid, tried-and-tested approach is to begin with the figures and their order of appearance, so we will discuss figures and their captions at length. For both Methods and Discussion, I will give some advice on what should generally be included and how to choose the order of presentation.

After developing a broad outline for each section—a plan of what should go in and in what order—the aim is to "quantise" the section contents into paragraphs. Paragraphs are the basic units of writing that form your text. The output of the outlining phase should be a paragraph-level outline of your paper, or at least of its most important parts.

The paragraph-level outline should, at the very least, contain the topics of each paragraph and the points that those paragraphs should make. It can also include notes, bulleted lists, sketches of arguments, and perhaps some references and citations. You might even want to sketch a sentence or two, but this can be done later as well—there is no need to write words yet.

The purpose of all this outlining is to help you to think more clearly. Outlining forces you to consider the big picture before spending time on details and makes it easier for you to follow your storyline. The paragraph-level outline should be an expanded version of the story that the abstract of the paper tells. Were you to write without a plan, you might find yourself in a situation where you simply end up somewhere, wondering where to insert the important points you originally had mind but that got lost along the way.

You don't need to be a perfectionist when it comes to the outline, and you don't need to decide every detail of every paragraph in advance. For some sections, outlining at the paragraph level might feel like overkill, and in these cases, you can simply come up with a sketch of what should go into the section.

Outlining the paper often results in the discovery of new connections and arguments, or you might find holes in your thinking that you need to fill before continuing. If so, great, do it—put your trust in the writing process and let it guide you. You might also find that outlining your paper throws up difficult decisions, but don't spend too much time dwelling on details: just skip that bit, and see if the process of writing solves your problem later.

As well as forcing you to look at the big picture first, outlining helps you to be more productive. When writing without an outline, it takes a lot of time to get started, and you have to read what you have already written just to remember what you were planning to say next. This cannot be done in a few minutes. Even half an hour might not be sufficient if the writing process has been interrupted for, say, a week or two. Therefore, if there is an unexpected gap in your schedule, it is hard to use this windfall of time for writing.

But if you have an outline, a plan, you can start writing easily and quickly at any time. With a paragraph-level outline, almost any available slot of time can be used for writing. Pick one of your paragraphs, look at your notes, and just write a sentence or two. Working in this way will help you to complete the first draft sooner than you'd think.

6

HOW TO WRITE THE INTRODUCTION, PART I: STRUCTURE

"You could find out most things, if you knew the right questions to ask. Even if you didn't, you could still find out a lot." — Iain M. Banks, The Player of Games

Much like the abstract, the introduction should be structured so that its scope follows an hourglass shape, from broad to narrow and then back to broad. But in contrast to the abstract, the hourglass of the introduction should be top-heavy, emphasising the context—the top of the hourglass—more than the resolution of the story.

Begin your introduction with a paragraph that sets up the broad context. This paragraph is important: it is the part of your paper that is most likely to be read (after the abstract of course). To get your reader curious enough to read on, the first paragraph should broadly point out a gap in knowledge that the paper aims to address.

After the broad context has been laid out, the scope of the introduction should become narrower. The next one to two paragraphs should bring the story to the confrontation phase: they should frame and motivate the research question of the paper. They should also cite relevant literature to provide context and to connect the paper to the streams of thought that together form your field of science.

Finally, after no more than a few paragraphs, the exact research question addressed by the paper should be explicitly and clearly stated. The sentence that reveals this question is the narrowest point of the top-heavy hourglass.

What happens next depends on the format of your paper.

For short papers and the letter format, the introduction has to be wrapped up rather quickly. A common approach is to summarise the main finding and explain how it was obtained in one or two paragraphs before moving on to the next section (Methods or Results).

For longer papers, it is common practice to provide a literature review addressing the state of the art. Rather than encompassing your whole field, this review just needs to provide an account of what others have done in the general vicinity of your research question. The literature review can be followed by a condensed account of your approach to the problem—your experiments, methods, or theoretical points of view—before you briefly present your main findings. Note, however, that there are field-specific "old-school" traditions of scientific writing where the results are not discussed in the introduction at all. Finally, the introduction of a long paper often ends with a map of the paper that gives an outline of what is to come: "In Section II, we will discuss..." and so on.

You'll learn a lot about how to approach your introduction if you pick some of your favourite papers and analyse their introductions, taking the time to understand the role of each paragraph. You will see that they almost always follow variations of the above arc. It is even possible to categorise papers by their type of introduction—how many paragraphs are there before the key problem is stated? One, two, or three? I've seen PNAS papers where the authors move from the broad context to the exact research question in one long paragraph, but this is an exception rather than the norm, and I would tend to split such long paragraphs. In any case, there is something of a formula, whose exact details depend on the format and length of the paper and the stylistic choices of the writer. The paper's topic and its familiarity to the readers of the journal also play a role: research questions that are obviously familiar to the intended audience do not require a lot of motivation, but new points of view or unexpected questions often do.

7
HOW TO WRITE THE INTRODUCTION, PART II: A FOUR-PARAGRAPH TEMPLATE

"The scientist is not a person who gives the right answers, he's one who asks the right questions." — Claude Lévi-Strauss

Following a good formula makes writing easier, so let's use the four plot phases (setup, confrontation, resolution, epilogue) as a starting point for building a template. I will now introduce you to a paragraph-level template that I often use when writing the Introduction.

This template is well suited to letters and short papers. It proceeds to the exact research question and to the main conclusion of the paper rather quickly—the whole introduction takes just four paragraphs. The first two paragraphs provide context and lead to the research question, and the last two discuss how the problem was solved and what the main conclusion of the paper is. The narrowest point of the hourglass is between these two blocks, immediately after the second paragraph.

It shouldn't be too difficult to expand this template for longer papers—just use more than one paragraph for each point, and add an outline of the paper at the end if you wish.

It is also perfectly possible to squeeze this template into two paragraphs: just merge the first two and the last two.

So, let's get going. *The first paragraph* provides context and background for your research by introducing the problem area and the knowledge gap that has led you to ask your research question. After the first paragraph, your reader should have a pretty good idea of what the paper is broadly speaking about—not necessarily the exact details but the broader topic.

Because this is the opening paragraph that leads the reader into your story, the first couple of sentences are crucially important. They not only determine what the first paragraph is about, but they also set the expectations for the whole paper. They are the beginning of the so-called "lede" (more on that in the next section) that draws the reader into your story. It is tempting to use the all-too-common opening in which you begin by explaining that your research topic has become important in recent years because of this and that. I've done this far too often, and I have promised myself that I will avoid it in the future. There are so many more exciting ways to begin your story! Say something powerful. Move directly to where the gaps in knowledge are.

After a strong beginning, you can continue the paragraph by giving a short overview of the state of the art, explaining what has already been discovered. The aim here is not, however, to provide an account of everything in the field—that's what review papers are for. Your overview should cite papers that provide context for your research problem and that connect your work to the broader progress of your field. By all means cite your own work too, if it is relevant to the problem—just don't overdo it.

This short literature survey of what is already known can fill the rest of the first paragraph. You can describe past research in chronological or topical order; what often works best is a funnel structure where you move closer and closer to your actual problem with every sentence.

You can make your first paragraph stronger by ending it with some contrast, for example, a sentence like "despite all this, we do not yet fully understand X" or "however, the role of Y remains an open question". The subject that provides contrast—the thing that is not yet understood—doesn't have to be the exact research question that your paper deals with. It could be something bigger, providing broader motivation for your question.

Equally, it is perfectly OK to structure the first paragraph so that each sentence simply adds detail and depth to the point made by the first sentence. The first paragraph does not have to end with a cliffhanger.

The second paragraph switches to a close-up point of view and zooms in on your particular research problem. Here, the plot advances from the setup phase to the confrontation. The second paragraph focuses on illustrating the gap in knowledge that has motivated your study. Your main aim is to get to your specific research question, which will be stated at the end of this paragraph.

Structurally, this paragraph forms a funnel from the broader motivation to the exact question.

Here are some devices to help you begin this paragraph. If the first paragraph concludes by contrasting knowledge with the lack of knowledge, your second paragraph can begin by addressing this contrast. For example, if at the end of the first paragraph you say that X is an open question, you can begin the second paragraph by explaining why this question is important, why hasn't it been solved, or what approaches might be feasible for solving it. If the first paragraph doesn't end with a question or a contrasting statement, you can launch the second paragraph with "However, ..." and add your contrast here. Your first few sentences should leave the reader aware of a question, a gap in knowledge, an issue to be solved.

Use the rest of the second paragraph to move from this broad issue to your specific research question, and tell the reader why answering that specific question will be worthwhile. If others have tried to tackle the question before you, tell the reader how they approached the problem, what they might have missed, and how your point of view relates to this existing body of knowledge. If you have come up with a unique question, tell the reader what is to be gained by solving it. Use carefully chosen citations to emphasise what is *not* known over what is known.

At the end of this paragraph, you should clearly and explicitly state the research question that your paper addresses.

At the beginning of *the third paragraph*, the point of view moves from what others have done to what you have done. The third paragraph describes how you approached the research question. In terms of plot, this paragraph advances from the confrontation towards the resolution. Now there is action, and things finally start happening. The narrowest point of the hourglass has just been passed (it is located precisely between paragraphs two and three).

The first sentence of the third paragraph should describe in concrete terms what you have done to answer the research question. For example: "To this end, we have carried out an experiment where..." or "In this paper, we investigate the relationship between X and Y with the help of..."

The rest of the third paragraph should shed more light on your approach. If you have designed and carried out an experiment to answer the research question that was made explicit in the second paragraph, describe this experiment here. If you have figured out a new theoretical approach to the problem, explain this approach. If you have collected and studied tons of data with the help of some new procedure, tell your reader about the data and the procedure.

If required—and if there is space—the third paragraph can be long; it can even be split into several paragraphs. But stick to the point: you are writing a paragraph for the Introduction, not for the Methods section. There will be time to fill in the details later.

Finally, *the fourth paragraph* of the template moves from your approach to your findings. It reveals the outcome of your work and briefly summarises your results. This paragraph should explain how you have answered the question that you stated earlier by sharing the new pieces of information that your work has revealed. It should also tell how the world has changed because of this new information: there is increased understanding, there are new ways of doing things, there are more dots to connect.

As the fourth paragraph is about the resolution of the story, it is something of a spoiler. But don't worry—everyone knows the ending already if they have read the abstract.

I often keep this paragraph fairly short for maximum effect. It summarises the key findings so that a busy reader can stop here, perhaps to return to the details later, but it leaves enough unsaid to whet the reader's appetite. Also, the brevity of the paragraph provides a nice contrast with the lengthy third paragraph; the shortness of the paragraph gives the impression of weight and importance.

HOW TO WRITE THE INTRODUCTION, PART III: THE LEDE

"Call me Ishmael. Some years ago—never mind how long precisely—having little or no money in my purse, and nothing particular to interest me on shore, I thought I would sail about a little and see the watery part of the world." — Herman Melville, Moby Dick

"In a hole in the ground there lived a hobbit." — J.R.R. Tolkien, The Hobbit

The first sentence of the first paragraph of any written piece of text is crucially important, as all writers of fiction know. Make it as strong as you can.

First impressions matter. The subset of potential readers who, having been lured in by the abstract, decide to have a closer look will then encounter the first sentence of the introduction. For them, this is another decision point: to read on or to stop.

The second-most important sentence is the second one,

and the third-most important sentence is the third one, and so on. A reader can choose to stop reading at any point, after any sentence. This means that the first sentence will be the most-read sentence of your paper. Your second sentence will be read by fewer readers than the first, and your third sentence will be read by fewer readers still, and so on (if we assume that readers do in fact begin at the beginning instead of jumping in at random points). You will lose readers sentence by sentence, whatever you do. This cannot be avoided.

The stronger the sentences, however, the lower the rate of attrition and the higher the chance that some readers will make it through to the last one. If you retain your readers' curiosity, they are more likely to stick around. Create contrast and tension for excitement. Use cliffhanger endings: pose a question, and then wait until the next sentence to answer it.

Journalists use the term *lede* for the first few sentences of a news story—presumably spelled that way rather than "lead" for historical reasons that involve mechanical typesetting.

The lede is the lead portion of a news story. It gives the reader the gist of the story and entices them to read the rest. While the lede should contain the essence of the story, it should not explain everything—it should raise questions so that the paragraphs that follow can satisfy the curiosity of the reader. Journalists even have their standard schemas for ledes, and one of the most commonly used is the inverted pyramid lede that compresses the who-what-where-when-why-how of a story into a single sentence or two. The journalist then adds depth and detail in the sentences that follow, in decreasing order of importance.

Let's have a look at some great openings and powerful first sentences.

The first sentence of Battiston *et al.*, PNAS 113, 10031 (2016), reads:

"Several years after the beginning of the so-called Great Recession, regulators warn that we still do not have a satisfactory framework to deal with too-big-to-fail institutions and with systemic events of distress in the financial system."

This is a powerful beginning that immediately points out the general problem addressed by the paper. It also makes the reader want to read on—who wouldn't want to know where this story is going?

Another example of a great opening, from Centola & Baronchelli, PNAS 112, 1989 (2015), reads:

"Social conventions are the foundation for social and economic life. However, it remains a central question in the social, behavioral, and cognitive sciences to understand how these patterns of collective behavior can emerge from seemingly arbitrary initial conditions."

Here, the problem that drives the research is clearly spelled out in the second sentence (however, the exact research question does not appear until the fourth paragraph). The introduction forms a funnel from the broad problem to the more detailed question.

Here is the beginning of the first paragraph of Altarelli *et al.*, Phys. Rev. Lett. 112, 118701 (2014):

"Tracing epidemic outbreaks in order to pin down their origin is a paramount problem in epidemiology. Compared to the pioneering work of John Snow on 1854 London's cholera hit [1], modern computational epidemiology can rely on accurate clinical data and on powerful computers to run large-scale simulations of stochastic compartment models. However, like most inverse epidemic problems, identifying the origin (or seed) of an epidemic outbreak remains a challenging problem..."

From the topic (tracing epidemic outbreaks) to the question (identifying the origin of an epidemic) in three sentences, with a brief historical detour of the you-know-nothing-John-Snow variety. This just works.

9
HOW TO WRITE THE MATERIALS AND METHODS

"It doesn't matter how beautiful your theory is. If it disagrees with experiment, it's wrong. In that simple statement is the key to science." — Richard Feynman

While much of this book has been about how to turn your results into an exciting story, we now need to put excitement aside for a while. I claimed earlier that papers are not just containers of information, but their Methods sections most certainly are. Their role is entirely utilitarian. So before we discuss form, let's discuss function.

A Methods section serves two purposes. First, it should let other researchers gauge whether your conclusions are justified and backed up by evidence; it should let them assess how credible your data are and how credible your analysis is. Second, it should allow other researchers to replicate your study and repeat whatever it is that you have done.

Unfortunately, as any experienced researcher knows, these goals are not always met. More often than not, the authors of a paper do not explain their procedures in enough detail, even if there is a supporting information document with an unrestricted page count. It happens all too often that when the reader attempts to really understand how the authors have arrived at their results, she has to give up because this information is simply not there.

Not being able to understand a paper's methods or replicate its pipeline leads to many problems. First, this contributes to the replication crisis and therefore erodes the very foundation of science, the scientific principle itself: only those results that can be replicated by others can be taken as facts. Second, it will be harder to sell your discovery to the scientific community if your fellow scientists cannot trust your findings because they do not understand how they were obtained. Third, if your pipeline—from data collection to analysis—contains new methods or ideas that others might be interested in, they will not be adopted by anyone unless they are clearly explained. This leads to many lost citations and to your work not being discovered. If you release data, someone will be able to use it to look for things that you didn't think of, and if you release your code and data analysis scripts, there will always be someone who needs them.

So please take replication and reuse seriously. Explain what you have done in as much detail as possible. Release your raw data. Release your intermediate results. Release your code. Reveal everything. Hide nothing. Be a good scientist. Don't be an evil scientist.

If you release everything that there is to release, you will probably need to use external repositories. Some journals, however, allow you to submit supplementary data and code files, to be published together with the article. If you are thinking of hosting the data and code yourself, keep in mind that we are talking about the scientific record here: your paper, your data, and your code should, in theory, be available forever. And forever is a mighty long time, as the late artist known as Prince once put it. It certainly is longer than the lifetime of the URL that points to your homepage on your university's server, or of the server daemon that runs on the Linux machine in your bedroom closet. So, no DIY here please—always use official data repositories such as Zenodo (www.zenodo.org). While even those might not last forever, they'll last longer than any self-hosted repository.

Let's return to the paper itself and move from function to form. The issue of where to describe materials, data, and methods depends on the journal, and there are many options. The top-tier journal style (think PNAS, Nature, and so on) is to have Materials and Methods as a separate section at the end of the article, as an appendix of sorts. In these journals, methods are only briefly described in the main text, and the reader is referred to the Materials and Methods appendix for details. This short letter format is all about the story, and technical details that would get in the way of the story are pushed aside. It may feel difficult to write this way because one cannot hide behind those technical details—there has to be a story. But beware of the dark side: referring the reader to a Materials and Methods section where only superficial details are given and where the reader is further referred to the Supplementary Information section that adds detail but still lacks essential information. Or where the limitations of the chosen methods are hidden in a subordinate clause on page 28 of the SI. This structure can make it dangerously easy to sweep something under the rug. Which is why it often happens.

If you are writing for one of these journals, resist the dark side: do not hide problems in the SI. Other than that, just strive for clarity in the Materials and Methods appendix. Typically, this section comprises independent subsections for different items with their own subheadings, so there is not much storytelling involved. In the main text, when talking about methods, describe their purpose, not their details. For example: "we measure the similarity of X and Y with the help of (insert name of fancy similarity measure), see Materials and Methods for details."

Then there is the style common to biomedical journals where Methods are described in all their detail straight after the Introduction. This makes it easier to report everything properly. It also makes it more difficult to hide problems, which is good. The downside is that being hit by several pages of painstakingly detailed method descriptions is something of a turn-off—the story suffers. While this cannot be entirely avoided, it helps if you remember to provide context: begin each subsection by reminding the reader why this dataset was collected, why this experiment was done, or why you are about to describe some mathematical methods. Often, this is no more difficult than simply saying that, for example, "to measure the similarity of X and Y, we need some well-behaved distance measure for probability distributions that..." and then describing the chosen measure.

The third way, commonly used, for example, by some journals of the American Physical Society, is to happily mix methods with results, explaining how things were done and what the outcome was without making a distinction between the two. In this case, things like experimental setups or data collection procedures may still be explained separately, but mathematical and statistical methods are typically described together with the results. In my view, this makes it easier to write a story that flows well. It is easier to motivate the methods by saying: "Next, we'll investigate X, and to do that, we need to do Y—here is how you do it, and look, here's the result." In the biomedical style, this connection is harder to make because the methods and results are separated, so one has to focus on making sure that the reader understands why the methods have been chosen and why the reader should understand their details.

Before concluding, let us return to being good versus being evil, and talk about discussing the limitations of your methods. All methods have limitations, as every scientist knows, and it is best to lay these out in the open. The Methods section is a good place for doing this. While limitations are often dealt with in the Discussion section, it is better to also address them immediately when the methods are introduced. This feels more honest to me. First, it feels a bit like cheating when the writer only mentions in the last paragraph of the paper that "by the way, we're not sure that things work the way we just told you they would". Second, it is easier for the writer to explain the limitations together with the methods. Third, it is also easier for the reader to understand the limitations and their implications if the details of the methods are fresh in her memory.

To summarise this chapter: when writing Materials and Methods, be a good scientist. Reveal everything and give the scientific community as much as you can, using data and code repositories in addition to your paper. Always motivate method descriptions, so that the reader doesn't have to wonder why you suddenly start explaining some procedure in detail. Discuss limitations of your methods openly.

10
HOW TO WRITE THE RESULTS, PART I: FIGURES

"We are lucky to live in an age in which we are still making discoveries." — Richard Feynman

A solid, tried-and-tested approach is to begin outlining the results section with the figures and their order of appearance. If you have followed the approach detailed in the first chapters of this book, your results already come with handily categorised labels (setup, confrontation, resolution, and epilogue). The main result is the resolution of the story, and the rest form parts of the setup, confrontation, or epilogue plot phases.

If your paper follows the standard Introduction-Methods-Results-Discussion structure, the Introduction doesn't usually contain any figures except for schematic diagrams of the setup category (which may also go into Methods). Figures and tables depicting your results will be placed in the Results section; within this section, the figures of the setup category come first and those of the epilogue come last.

So you've already done much of the work before you

have even begun outlining the Results section—the respective plot phases of the figures mostly determine their order. All that remains is to choose the order in which the results and schematic diagrams are shown within each category: which figure leads to the next?

The order of figures should tell a clear story so that each figure builds on the previous ones. You can use a multi-panel figure to tell a self-contained part of your story, forming a miniature story arc. You can combine, for example, a schematic diagram that explains your experiment (setup), some basic statistics of your data (setup), and a result plot that contains an unexpected finding (confrontation). This three-to-four-panel mini-story figure is a technique I often employ in letter-format papers where the story has to move fast and get to the point quickly. The figure already brings the story close to the resolution and the key result.

When choosing the order of presentation, focus on making your point as convincingly as possible and in the most exciting manner. This doesn't necessarily have anything to do with historical accuracy—there is no rule about presenting your findings in the same order as they emerged. A paper is not a historical account of your research: it is a story that makes a point.

When you have chosen the order of your figures, write a draft version of the caption for each figure. You may need to revise the captions later, so don't worry about making them perfect. But try to ensure the captions are self-contained enough for a hasty reader to be able to grasp the main point of your paper just by glancing through your figures. In fact, this is exactly what a great many readers do—they just skim. This is also what the editor of your journal will do before deciding whether your paper is worth a closer look or deserves to be rejected outright. So help your readers and make skimming easier by writing easy-to-follow, self-contained captions.

How long to make the captions depends on your journal. In some of the letter-format top-tier journals, captions tend to be quite long and contain almost the entire story. In the lesser journals where us mere mortals are published, figures are usually discussed at length in the main text too, and therefore the captions can be shorter—but resist making them too short. Always make sure that your captions reflect what the reader should learn from looking at the figures. A caption that only states "here we see Y plotted as a function of X" is useless. This caption is entirely redundant if you have already labeled the axes of your plot (and if you haven't, repent for your sins, and label them NOW). Always tell the reader what she should see in the figure: how should the plot be interpreted? What is the message of the figure?

Because much of your story will be told by figures, let's talk about figure quality. Figures are tremendously important because those who only skim through the paper won't see much else. And, I repeat, this means the majority of your readers. Figures make the first impression, and first impressions matter: good-looking figures may convert skimmers to readers. Clear, high-quality figures that look professional demonstrate that considerable effort has been put into the paper, rendering the reader more likely to trust its contents. Amateurish-looking figures with a colour scheme straight out of the 1995 version of PowerPoint will leave the reader wondering if the results are of the same dubious quality.

So make sure that your figures look good. To do this, learn the ropes of whatever program you use to generate your figures, whether it is a Python or R library or a stand-alone piece of software (like Gnuplot, which has been around since the dawn of man and will probably outlast even cockroaches once mankind has been wiped out). In particular, learn how to change fonts, how to increase or decrease font size, and how to use proper LaTeX-style fonts wherever appropriate. Learn how to choose and manipulate colours, colour schemes, symbols, and shadings. Learn how to produce figures of particular dimensions. Learn to match figure size to your target journal's column width—not having to scale the figures takes some of the guesswork out of choosing font sizes (see below).

Learn to use a vector graphics editor to post-process your figures (and learn the difference between bitmap images, which are made of pixels, and scalable vector graphics, which are made of lines and arcs and Bezier curves). At the time of writing, the industry standard (design industry, that is) would be Adobe Illustrator, but there are many free alternatives such as Inkscape. With a vector graphics editor, it is easy to assemble multi-panel figures that combine schematic diagrams (drawn with the same editor) with plots saved in vector formats (PDF, SVG). You can also add text, arrows, indicators, and so on, and you can retouch your plots, change line widths, and alter the colours of symbols or their overall appearance. Often, this is much faster than trying to get everything right when producing the plots with your data analysis or plotting software.

A few words on the layout: always align things. Nothing says "I am being careless" more clearly than subplots and schematics that are not lined up neatly—it takes just a few seconds to do this. Use white space properly: leave enough white space so that things can breathe, but don't leave so much white space that the figures look barren.

Discussing data visualisation at length is beyond the scope of this book, but I'd still like to mention a few things.

Let's start with colour: pay attention to your colour schemes. For plot symbols, there are nicer, more informative schemes than the pure-RGB red, green, and blue symbols that some programs use by default. Additionally, your reader might have a hard time distinguishing between red and green. For maximal clarity, always use different symbols AND colours for different curves. If you want a personalised colour scheme, search the web for colour scheme generators (you have already learned how to set hexadecimal colour values in your program, right?). For things like heat maps, pay attention to the colours you use: make sure that they don't artificially highlight some range of values (for example, by displaying a range of data points in striking red even though those points are not that different from the others). In all cases, use colours consistently throughout your figures. If red and blue are categorical indicators of two different data sets in one plot, don't use red and blue to indicate high and low values in a heat map elsewhere in your paper. Reserve red and blue for the two data sets and always use them this way. Likewise, if you use a colour map with a gradient from low to high values, reserve its colours for this purpose alone.

Next, think carefully about labels and fonts. Always label your axes. This is self-evident but worth mentioning explicitly. Forgetting to label the axes of a plot *should* feel like forgetting to get dressed when leaving for work in the morning—and yet it still happens all too often. So, I repeat: label your axes, period. At all stages of your work, even if the plot is just a draft for your own eyes. And when labelling, make sure that the fonts you use remain large enough when the figure is scaled to its intended size. Not paying attention to font size is a very common problem among beginners, but there are many published papers where you need a magnifying glass to understand what is going on in the figures. I suspect this problem is rooted in the defaults of some commonly used software packages, as default font sizes are almost always tiny. I've rarely (if ever) seen plots with annoyingly large fonts, so if in doubt, double your font size.

Finally, a few words about "having an eye for design". While beautiful and impressive figures seem to come more naturally to some, every student can learn to produce good-looking visuals. I've heard many people say "I cannot draw, and therefore my figures look ugly", but—as with any skill—it just takes time and patience; you don't need to go to art school to learn the essentials. Just as the key to illustration is learning to observe things carefully and analytically, the secret to producing great-looking figures is understanding how they should look. This is best learned by imitation. The next time you have a plot that you are not entirely satisfied with, look up a similar figure in a journal article that you like. Look at the two figures side by side, and try to spot the differences in composition, colours, fonts, line widths, and so on. Then modify your figure so that it resembles the reference figure more, and keep on modifying it until you are satisfied with the outcome.

11
HOW TO WRITE THE RESULTS, PART II: TEXT

"The scientist only imposes two things, namely truth and sincerity, imposes them upon himself and upon other scientists."
— Erwin Schrödinger

The arc of the results section follows the order of the figures, from the setup to the confrontation and from the resolution to the epilogue. The results of the first phase (setup) introduce the reader to your data. They make your final conclusions credible by showing what is in your data and letting the reader gauge whether it looks OK. The results of the second phase (confrontation) point out that there is something surprising that needs to be sorted out. The results of the third phase present the resolution of the problem, and the results of the epilogue phase show what follows from the main finding and why that matters. They would not necessarily work as stand-alone results.

Make sure that the reader can easily follow the arc. She should always know where she is and where she is heading next. One way of making sure the reader is on the map is to further divide your results into subcategories, discussed in their own subsections that come with informative headings. You can develop subsection headings by condensing each result into a single short sentence and then using this sentence as your heading. This way, each result gets its own subsection where it can be explained in detail. Note here that "a result" does not necessarily mean a single plot or figure but rather a conclusion that may be based on several pieces of evidence.

With your results section organised in this way, the reader can get a quick overview of the whole section by scanning the subsection headings and the figure captions. Again, remember that most readers just skim.

The above technique is an example of so-called signposting, where the whole paper is made more accessible by covering it with signposts that tell the reader where she is. Clear section headings help, as do clear figure captions whose first sentence should explain what the figure is about. Clearly formulated key phrases are also very useful.

In your Results section, the most important signposts are the subsection headings, the first sentences of figure captions, and something we haven't discussed yet: the first sentence of each results subsection. These sentences should motivate and provide background for the subsection: why was this experiment conducted? Why was the analysis done that led to this result, and why are we discussing this result? A useful construct to launch the first sentence of a results subsection is "To understand X, we measured Y..."—or something similar. Even though the motivation has already been mentioned in the Introduction and possibly in the Methods as well, it is still good practice to begin with a sentence that reminds the reader of why she is reading this part (unless the paper is a very short letter and the introduction appears just a few paragraphs earlier).

So, what else is there in a results subsection? Well, results, of course. But there are several layers here.

The lowest layer contains your "pure" results. These are essentially just data: you measured X, here's what you got. You computed Y, here is a table. The second layer contains direct and unambiguous interpretations of these data: "the distribution of X measured under condition A clearly has a lower mean than when measured under condition B," or "Y grows faster as a function of time than Z." And so on. While such statements may already contain the main conclusion, an additional layer of interpretation is usually required to give meaning to the findings.

The above layers form a logical order of presentation for each of your results subsections. Begin the subsection with motivation and background. Then briefly tell the reader what you have done, either by referring the reader to the Methods section or, if you are writing mixed results and methods, by presenting your methods here. Then explain the results that you have obtained and talk about how you interpret them. Make it clear which layer you are talking about: what is an indisputable fact, what is interpretation, and what is speculation. Use signposts for this. For example, saying "these results can be interpreted as follows" helps the reader to understand that there might be other ways of interpreting the results. Do not exaggerate or over-generalise: if there are limitations that haven't been addressed already (say, in the Methods section), be open about them.

There are traditions in some disciplines, such as certain biomedical fields, where the Results section is strictly about results. In this fundamentalist interpretation, the Results section should only contain the first layer (pure data) and perhaps some of the second layer ("these data have a lower mean than those data"). The third layer—the interpretation of the results—should then be left for the Discussion section and not even be mentioned in the Results. To me, this is borderline insane (and may have been invented by the same evil people who run journals that send papers to referees with the figures separated from the text and the captions separated from the figures). Why make the reader's life difficult by forcing her to jump back and forth between the Results and the Discussion? If you work in one of those disciplines where this is the norm, it may still be possible to have a separate discussion section that directly follows each result (or subsection of Results). If not, I guess you'll just have to obey the rules, or at least look like you are obeying the rules. But don't do it willingly: always try to sneakily insert at least one sentence that explains your findings. Rejoice if the editor and the referees don't notice it!

Finally, before we move on to the Discussion section, I have one more trick to share. It is common to begin a paragraph in the results section with "In Figure 1, we see that..." Next time you read a paper, try to be conscious of your own response to this. What do you do? Do you directly jump to Figure 1 and look at it and then try to get back to the middle of the sentence that you were just reading? Can you resist the impulse? Even if you can, you still FEEL the impulse, don't you? It is impossible not to feel it. And the impulse makes it harder for you to follow the sentence to its conclusion. So always refer the reader to the figure last, not first: FINISH the sentence with "as we see in Figure 1" or something similar. This way, when the reader arrives at your mental hyperlink, she has already read and processed your sentence and knows what to look for in Figure 1.

12

HOW TO WRITE THE DISCUSSION

"Before the result of a measurement can be used, it must be interpreted—nature's answer must be understood properly." — Max Planck

Time to wrap up the paper! You have presented your research question, its greater role in the universe, and your findings. Now let's close the circle and discuss your answers and the questions that still remain, or the questions that you have just discovered. This is what the Discussion section is for.

Before we get started, a note for those poor souls who work in fields where you are not supposed to discuss your results in the Results section at all (see the previous chapter): in this case, the Results section just contains data, and therefore the Discussion section usually has two parts, Discussion and General Discussion. The first part, Discussion, is about interpreting each of your results, which I have already covered in the previous chapter. The second part, General Discussion, is what we will cover next, and I'll just refer to it as Discussion from now on. Sorry for the confusion (not my fault).

And while we are at it, what's the difference between Discussion and Conclusion? The Conclusion is typically fairly short—it is almost like an echo of the abstract. It is a summary of what you have done and adds no new ideas or thoughts. In contrast, the Discussion section can introduce new points and insights and show how your findings relate to the broader issue at hand.

In the Discussion section, you should remind the reader of your research question and provide a synthesis of your results. You should present the answer to your question and show how it contributes to solving the broader problem that you painted in the Abstract and the Introduction. You should suggest further research based on your results. You should also contextualise your results (and your research problem) within the literature—this is a good opportunity to cite more papers that illustrate where your results fit in and what they mean. It is quite common to cite papers in the Discussion that were not cited in the Introduction: the Introduction and the Discussion are usually the sections of the paper that feature the most citations.

When discussing the significance of your results, avoid over-generalisation even though everyone else does it and it creates catchy headlines for journals and the media. Don't exaggerate. Be an honest scientist. It will pay off in the long run. You can also turn any generalisability issues into questions: instead of claiming that your observation is generally valid, ask if it can be replicated under different conditions or in different sets of data. You never know, one of your readers might take up the challenge.

A good Discussion section contains plenty of questions

in addition to answers. These questions may be concerned with the remaining knowledge gap: what else needs to be found out on top of your results to solve the broader problem, or to further research in the area that your work has addressed? Perhaps, as often happens, there are genuinely new questions now—questions that your work has revealed.

While it is common to discuss the limitations of your work in the Discussion section, this should not be the first time they are mentioned. Let the reader know about them earlier—be open from the very beginning. All technical limitations should be discussed together with your methods. You should also be clear about what you can and cannot conclude from your results in the Results section.

When addressing limitations, you should tread carefully: there is a difference between being honest and making it sound like your study is flawed. Joshua Schimel's *Writing Science* introduces a great principle: say "but, yes" instead of "yes, but". Instead of saying that your results would be even clearer if your experimental setup had a higher resolution, say that even though the resolution of your experimental setup is limited, your results are nevertheless quite convincing. The latter has a much more positive ring to it even though both sentences have the same information content. Don't make it sound like there is something wrong with your work—and if there is, fix it before writing your paper.

In general, there is plenty of freedom to choose how you shape the Discussion section. One common approach is to begin to remind the reader of the broader knowledge gap and the specific research question of the paper. You can then proceed through the results one by one, perhaps grouping them to make certain points, and summing up all evidence before arriving at the final conclusion.

Alternatively, you can begin the Discussion by recapping your research question and your main result. This is the inverted pyramid style, where you begin with the key points —the question and the answer—and then add the details and different points of view, which often appear in decreasing order of importance.

Whatever structure you choose, the last paragraph of the Discussion section is very important. It contains the last words of your paper, your final thoughts before getting into back matter such as Acknowledgements. Do not waste those words! Endings are power positions, and the last words of your paper will be remembered by your readers (or at least by those who made it this far and the skimmers who jumped directly to the end). So always end on a high note. If the reader has been with you this far, reward her at the end.

One good way of ending the Discussion section—and the paper—is to write a short paragraph that recaps your conclusion and the significance of your work. You can even signpost this to the reader and begin the paragraph with the words "in conclusion, we have shown that..." This paragraph can be written in the style of a news lede. What did you find out and why? What does it mean? What are its broader implications? How has your work concretely contributed to the big picture? Where have you arrived at from your starting point? How has the world changed because of your results? This paragraph closes the circle and resonates with your Introduction.

There are some all-too-common ways of ending a paper with a whimper instead of a bang. One such way is to focus on the limitations of your work in a negative light. Another common problem is ending the paper with a vague statement. "Further research is needed" is a platitude and a vague one at that. Further research is *always* needed! Instead, go for something like: "Because of the results of this paper, we are now in a position to tackle problem X with method Y, bringing us closer to the ultimate goal of Z." This is far more concrete and memorable.

PART III
WORDS

13

HOW DOES YOUR READER READ?

"Words are to sentences what atoms are to molecules: the basic building blocks that control structure and function. If we extend that analogy, paragraphs become cells: the fundamental unit of life. A cell gains life from its structure, a structure that creates internal cohesion and external connection, allowing it to function as part of a larger organism." — Joshua Schimel, Writing Science

Finally, after all this planning and outlining, it is time to start filling in the blanks. It is time to write words that form sentences that form paragraphs that form sections that form your story.

Because of the top-down approach that has brought you here, coming up with words and sentences should now be easier than if you had simply started with a blank page. You should already have an outline of the paper and a plan for each paragraph—now you just need to turn those notes into full sentences!

How do we write effective sentences? From the point of view of the reader, the best sentences are those that are easy to understand and say all the right things in the right order. The reader should always be able to anticipate what comes next. From the point of view of the writer, feeling empathy for the reader is useful. A good writer tries to look through the reader's eyes, taking her hand and guiding her through the text, making her job as easy as possible.

To act as a guide through the text, the writer has to gently manipulate the short-term memory of the reader. It has been argued that one's short-term memory can only hold seven things—such as digits—at a given time. When it comes to concepts that are more complex than digits, even seven sounds like an awful lot to me. There is only so much anyone can hold in their head.

When a piece of text feels too hard to grasp, this is not necessarily because the ideas therein are difficult. It might be due to a problem with the sentences whose job it is to deliver those ideas and to make them comprehensible. Often, the order of presentation is at the root of the problem. The reader has to struggle if she has to keep too many concepts in her short-term memory without knowing how they are connected. Bad writing randomly jumps from one thought to another. This creates a confusion of thoughts and ideas without giving a clue as to how they relate to one another.

What is the right order of things, then? What should come first in a sentence or in a paragraph? To answer this question, we have to consider how the reader processes information—how the reader reads. So, let's think of the reader as an automaton of sorts, with three types of memories that you can manipulate. Each type of memory plays a different role in interpreting and understanding the words, sentences, and paragraphs that the reader encounters.

Long-term memory contains the key concepts that are

required for understanding the paper. This memory is initialised in the Introduction, and populated with more items as the reader reads on. Whenever new concepts become established in the paper, they are added to this memory.

Medium-term memory holds the concepts that are essential for connecting the dots in front of the reader—for making sense of things in the current paragraph. In particular, this memory contains the current paragraph's topic (or the reader's interpretation of it). Not all concepts remain in this memory for long: many are flushed out at the end of each paragraph. If you are familiar with time series analysis, you can think of this memory as a sliding window of sorts.

Short-term memory holds concepts that are required for making sense of the current sentence: what its subject is, what its verb is, and what its object is, or in other words, who does what to whom. In addition to grammar, this short-term memory is essential for relating the concepts encountered in the sentence to one another. As the reader reads, the words and concepts of a sentence are placed in this memory in their order of appearance. This memory is emptied at the end of each sentence.

How do these different types of memory work? When the reader reads, she consumes the words from the left to the right and interprets them with the help of all three. The short-term memory helps to parse the current sentence and link its words to one another. The medium-term and long-term memory determine how words and concepts are interpreted; they help to extract meaning from the sentences.

If all goes well and the reader understands the words in front of her, the concepts that those words hold are first put into the short-term memory, to be flushed out at the end of the sentence once the reader has understood its meaning. At that point, relevant higher-level concepts formed by the words in the short-term memory are moved to the medium-term memory, to be used in deciphering the rest of the paragraph. At the end of each paragraph, this medium-term memory is given a vigorous shaking so that unnecessary baggage falls out. Those concepts that stay and prove to be useful may transcend to the long-term memory.

But if the reader encounters something that is inexplicable—something that doesn't match with anything in any memory—an error occurs and the reader is lost. Or, worse, if the reader parses the current sentence using a wrong concept, she can become entirely derailed.

Because of the way this parsing automaton operates, the first words of a sentence are tremendously important, and so is the first sentence of a paragraph.

The first words of a sentence initialise the short-term memory, determining how the rest of the sentence is understood. They define the topic of the sentence. As the reader reads on, the words that she encounters are interpreted in a way that makes sense in light of this topic. The more words accumulate in the short-term memory, the more the reader has to struggle to connect the words to one another to make sense of the sentence. In particular, if the sentence is constructed in such a way that its meaning hinges on its final word(s), it becomes taxing to read and difficult to comprehend.

So based on our model of the reader's short-term memory, the perfect sentence begins with words that clearly define its topic. The perfect sentence is also short and compact. It helps if the order of words allows the reader to parse most of the sentence up to the current word. Place familiar words and concepts at the beginning of the sentence, and use them to explain anything that might be unfamiliar later on in the sentence.

A similar logic applies to paragraphs. The first sentence of a paragraph establishes its topic, which is stored the medium-term memory and used to interpret the paragraph. So choose the first sentence well, and make it sure that the reader can immediately understand what the paragraph is about. The first sentence sets the expectations of the reader and determines how she will attempt to interpret everything that follows.

One way to make the topic of the paragraph clear is to begin with a phrase that acts as a signpost, telling the reader where the paragraph is going. Examples include "To measure how X depends on Y, we constructed an elaborate apparatus..." and "In conclusion, in this paper, we have shown that..." When the reader knows what to expect, interpreting the rest of the paragraph is easier.

Once you have established the topic of the paragraph, always stick to it. An unexpected, unconnected sentence breaks the flow and leaves the reader baffled. But a sentence that makes sense in light of the topic will be interpreted properly, and it will set expectations for the sentence that follows.

It's not only beginnings that are important—endings matter, too. The end of a paragraph and the end of a sentence signal a break, so the reader has more time to think about the last few words. This means that the last words carry extra weight in the mind of the reader, almost unconsciously. You can take advantage of this by placing things that you want to emphasise in these stress positions.

14

HOW TO WRITE YOUR FIRST DRAFT

"To write is human, to edit is divine." — Stephen King

The best and most productive writers do not write perfect first drafts. The best and most productive writers write crappy first drafts, and they do this as quickly as possible. Then they edit, revise, and polish their drafts until they are no longer crappy (and no longer drafts). Or until the deadline makes them stop—whichever comes first.

This is what you should do with your paper, too: write the first draft quickly, and then edit, revise, polish, rinse, and repeat until you are satisfied with the outcome. Or until the deadline arrives.

If you have followed the approach of this book, you are now at the point where you are ready to write your first draft. You have a story, you have a structure, and you have a plan for each section and each paragraph. If you have read the previous chapter, you have an idea about how to organise the building blocks of paragraphs and sentences.

This is all you need to write your first draft. Later I will provide lots of tips for editing your draft.

For the time being, write a complete first draft quickly and dirtily. Forget perfection. Let your first draft be human, let it be imperfect, let it be crappy! Writing and then polishing a crappy first draft is much, much faster than agonising over every word and sentence. Making perfect choices takes forever. Spend that time on editing and revising instead.

Now that you finally have to produce some text, this is where the pain of writing typically hits you. Coming up with plans and storylines can be fun, but writing rarely is. Writing is hard work, and a first draft is particularly difficult because it is hard not to be self-conscious and hard not to let your inner critic stop you mid-sentence.

This prospect can feel overwhelming, especially for beginners. How can we ease this pain? How can we write all the words that need to be written in order to produce a paper? Let's discuss some possible solutions.

First and foremost, make the initial draft your own (crappy) little secret. It is not for your supervisor or your co-authors to see—it is for no one else's eyes but yours. It serves as raw material for editing only. When your supervisor asks you for the first draft, give her your second draft instead (by all means, call it the first!). Keeping your first draft private should make you less self-conscious, at least in theory: no one except you will ever see it, so it doesn't matter that it's imperfect.

Aim to produce more text than you need. Just let the words come! At this stage, it's fine to have sentences that are too long, it's fine to repeat yourself, and it's fine to explain the same thing over and over again in different words. If you are writing, say, one of those four-page letter-style papers with restricted word counts, do not think about the length at all. Just write. Cutting text is easier than producing it, and the editing phase easily reduces the length of your text by 10-30 per cent. In my experience, the more the better.

In order to be productive, schedule your writing time and stick to it. Never wait for inspiration to strike because it rarely strikes those who just sit there waiting. The muses dislike idleness: they tend to show up when you are already engaged in work. Just sit down, put your phone on silent, remove all clutter from your screen, shut down your internet access, and do it. Write. A good target is something like 30-45 minutes of uninterrupted writing, followed by a break. For a really good day's work, four to five sessions are enough. Just keep on doing this daily until you find yourself at the end of your draft.

If you get stuck, try changing the way you write. Take a pen and a notepad, and walk away from the computer. Sit down somewhere, get a cup of decent coffee, and sketch your sentences on paper. Try to write as if you were making lecture notes or just jotting down ideas. Once unstuck, go back to your computer and use the material in your notes to continue. Or, instead of a notepad, try dictation, or perhaps go for a walk, and play out imagined conversations in your head where you explain to someone whatever it is that you are supposed to be writing.

If you have a problem with letting go and find it hard to make progress because of that nasty self-conscious voice in the back of your head, one option is to try the Morning Pages technique. This technique was introduced by Julia Cameron in her book, *The Artist's Way,* as a tool for artists to overcome whatever it is that holds them back. The Morning Pages technique uses stream-of-consciousness writing to provide desensitisation through repeated exposure. Take a pen and a journal each morning and write three whole, longhand pages about your research, filling them with anything that comes into your mind. See where this leads you! This may feel difficult at first, but just keep on writing.

If you feel that it is impossible to get some part of your text just right, this is often a sign—a message from yourself. When you are stuck with a paragraph that just won't yield, stop trying to force it. Instead, ask yourself: why is this so difficult? Search your feelings. What would make that paragraph easy to write, what are you missing? Often, you will notice that you are not faced with a writing problem at all—rather, you are missing some important piece of understanding. Perhaps your result is not clear after all, or you have not thought enough about some tricky issue, and that is why you cannot express it in words. If so, take a timeout, and look for understanding first; the words will come more easily when you have found it.

15

HOW TO EDIT YOUR FIRST DRAFT

"Write with the door closed, rewrite with the door open." — Stephen King

If you have followed the advice in the last chapter, you should now be the proud owner of a crappy first draft —a draft that serves as raw material and is for your eyes only.

Now it is time for you to put on another hat and play a different role. It is time to look at your draft critically and to examine each and every sentence and paragraph ruthlessly, so that you can cut out anything that doesn't pull its weight.

Before you do that, however, it might be a good idea to take a step back from your paper because a fresh pair of eyes can better spot what needs to be done. If possible, do something else for a few days and then return to your paper.

The process of editing and revising is iterative, and it can take many rounds: my most-cited paper was at version number 27 or so when it was finally submitted. This may sound a bit excessive, but hey, it worked! You don't always need to go to that length; just be sure to complete several rounds of revisions, first alone and then with the help of your co-authors and/or your supervisor.

A top-down approach, like the one we took when writing our draft, will work best when revising it. Begin by addressing broader issues—read the draft quickly, without getting stuck on sentences, words, or other nitty-gritty details. Focus on story and structure: is the story logical and exciting? Are concepts introduced in the right order? Are there sections that are too long or lacking in detail? I'll provide you with handy checklist that you can follow in Chapter 16.

The answers to the above questions may result in the need to "remix" the paper: to shuffle its contents around, to reorder things, and to completely rewrite some sections. This is normal, and just do it if you feel the need. Then, repeat the top-level analysis of your draft and see if you can think of ways to improve the text further. If the answer is yes, do it. Repeat this loop until you are happy with the outcome and satisfied with the overall structure and flow of your paper.

At this stage, you may even feel like returning to your research, perhaps to look for new results that could strengthen your conclusion. If so and if there is time, then by all means do it, but please remember that your research must stop at some point. There will always be new findings just around the corner—leave some of them for the next paper.

When you are happy with the structure, you can focus on paragraphs and sentences with clarity and readability in mind. Again, I'll provide you with a checklist (see Chapter 17). Check your figures as well (see Chapter 10).

Finally, when everything seems to be in place, do a shortening edit with the aim of removing clutter and superfluous words. Make every sentence shorter that can be made shorter. Remove all adjectives, unless they are completely necessary. Remove all repetition. Remove words that exaggerate things because you will sound more confident without them. Remove every instance of the word "very" because you never need it.

When you are ready to show your improved draft to others, you can apply a technique that my research group has borrowed from the software industry: Extreme Editing.

In the software industry, "Extreme Programming" is one of the fashionable agile techniques, part of which involves programming in pairs. So, edit in pairs! Or, if there are more co-authors, involve as many of them as possible. Force your PhD supervisor to reserve several hours of uninterrupted quality time with you and your draft. You can argue that this is a good investment because a co-editing session takes less time than several rounds of traditional red-pen commenting.

This is how Extreme Editing works: go to a meeting room with a large enough screen and open the draft on the screen, zooming in so that everyone can easily read the text. Then, go through the text together, paragraph by paragraph and sentence by sentence. Be critical of each word and each sentence; look for sentences that are unclear or those that could be misunderstood. Try to find ways of reducing clutter and shortening sentences. Trim the fat wherever possible. In my group, we jokingly keep a tally of points scored for every word removed. The winner is the one who has most ruthlessly killed the largest number of words that just tagged along, doing no real service to the text.

In the following two chapters, I will present some detailed tips for revising your draft, first focusing on the level of meaning and structure, and then on the level of sentences.

16

TIPS FOR REVISING CONTENT AND STRUCTURE

"Rewriting is the essence of writing well." — William Zinsser, On Writing Well

As discussed in the previous chapter, when editing your paper's first draft, my suggestion is to do two passes. During the first pass, you should focus on structure and content: does the draft contain the right things in the right order? Then, once you are happy with the big picture, do a second pass that focuses on words and sentences.

Here is a checklist for the first pass:

- Check the abstract and make sure that it follows the hourglass structure: broad context, narrower context, your research question, your result, your result's implications on your (sub)field of science, the broader implications of your result.
- Check that your paper is focused. Make sure it is consistent with the choices you made at the outset about the point of the paper and its key

conclusion. Leave out results that are not required for supporting the key conclusion, or safely tuck them away in the supplementary information document. When editing, if you feel that your paper loses its focus at some point, take a step back and then try to address the issue; if needs be, see my advice on remixing the paper in the previous section.

- Check that there is a clear question and a clear answer. A good paper states and then solves a problem; your results are only meaningful if they solve a meaningful problem. Remember that your paper is neither an account of your work nor a lab diary; it should be a story of an important problem and its solution. Emphasise the problem, both in the Introduction—where it should really stand out—and in the Results and Discussion sections. Show the reader how each result contributes to solving the problem, and what the implications of solving the problem are.
- Check that the figures tell your story. If you just glance through the figures and skim their captions, do you get the point of the paper? If not, go back and work on them.
- Check that you have provided all the necessary information for the reader to be able to replicate your results. Verify that your Methods section (and the supplementary sections if you have any) contains everything that the reader needs to know. Also check that you have provided links to your code and your data.
- Check that the Discussion section contains questions, not just answers.

- Check that the paper ends with something concrete and worth remembering. Endings have power. Do not waste this power.
- Check that you have provided enough background information: your reader does not know what you know. Assuming that your reader knows much more than you and consequently omitting necessary background information is a very common problem among students. A typical example would be a Methods section that directly launches into what you have done without first explaining why. While it is evident to you that to get from A to B you need to do X, this is probably far less obvious to the reader. Never assume that the reader knows your motivation or the details of every method you. Tell her. Many students seem to think that they know a little while everyone else knows a lot. It is only later in their careers when they realise that no one really knows that much! Besides, your readers will include people from adjacent (sub)fields and readers who are just learning the tricks of the trade. Ask a colleague who works on something slightly different to you to be your test reader. Ask her which parts of the text are hard to follow, and revise accordingly.
- Make sure that your writing is not confusing—your reader shouldn't have to work too hard. Feel empathy for your reader and try to get inside her head, assuming that she knows nothing or very little. Gently guide your reader through your text. Steadily bring the reader from result to result, from paragraph to paragraph, and from sentence

to sentence. Never leave it to the reader to connect the dots—always connect the dots for her. Err on the side of caution. Papers where things have been over-explained are rare (if they exist at all), but papers that are too difficult to follow are frustratingly common.

- Check that your nomenclature and notation are consistent. Using an outside reader as a guinea pig is recommended here, as this problem may be hard for you to spot because you have been immersed in the world of your paper. Problems with naming things are notoriously tricky to detect; problems with notation are slightly easier but can still prove difficult. There are often a number of concepts floating around when you are doing research and conceptualising the paper, and the very same things can take many names in your thinking. Writers of fiction are allowed to use synonyms for variation, but science should be precise: do not use multiple terms for the same thing in the final version of your paper. For example, it may be apparent to you that the thing you call "the weight matrix" is the same thing that was called "the correlation matrix" in the previous paragraph, but your reader will quickly get confused.

17
TIPS FOR EDITING SENTENCES

"Don't use big words. They mean so little." — Oscar Wilde

After you have completed your first pass and when you are happy with the overall shape of your paper, it is time to polish the words, sentences, and paragraphs. Once again, your overall goal should be simplicity, clarity, and readability. You can achieve this goal by cutting out everything that is not necessary. Condense and straighten your sentences so that they are short, to the point, and easy to follow. When editing your sentences, pay attention to the following:

- Ensure that each sentence in a paragraph belongs to that paragraph: the first sentence defines the topic of the paragraph, and the rest stick to that topic. If a sentence goes off in a tangential direction, delete it or move it elsewhere; if the paragraph is long and its topic appears to change along the way, split the paragraph into two or more.

- Check that your sentences follow one another logically and that there are no jarring, abrupt changes in direction. Tie your sentences together with transitional words. Begin your sentences by addressing the words or concepts that finished the previous sentence, or use conjunctions (however, in addition, to the contrary, and so on).
- Check sentence length; aim for short, precise sentences. Try to make every sentence shorter by rephrasing ideas using fewer words and by cutting out words that don't contribute anything. Those lazy words include repetitions and unnecessary adverbs and adjectives. Also look out for wordy expressions involving the passive voice or nominalisations of verbs (more on this in a moment). Rejoice with every word that you cut! If your sentence still feels too long after tightening, look for ways of splitting it. Long sentences are taxing to read because the reader has to hold a lot of words in her short-term memory. You don't need to count words to detect sentences that are too long—you can spot them visually (if a sentence spans several lines, it is too long) or, better, by reading your text aloud. Wherever you stumble or run out of breath, you have a problem.
- Establish meaning early in the sentence, and keep your subject and verb close. If the first words explain who does what, it becomes easier to decipher the rest of the sentence. If the main point of the sentence is made clear from the outset, even wordy sentences can be comprehensible. But if the meaning of your

sentence is only unlocked by its 27th word, the long and winding road to get there will be littered with the remains of readers who have perished from sheer mental exhaustion.

- Use active voice and avoid the passive (for exceptions to this rule, read on). There is a long tradition of using the passive voice in scientific literature, probably because the passive voice sounds more distanced and impersonal—it sounds more "academic". But because we usually write about abstract concepts in the first place, there is no need to make them any more abstract, impersonal, or distant! So avoid the passive wherever you can. How to spot the passive when editing? Easy—if your sentence ends with the actor ("by X"), or if you can insert "by zombies" after the verb without violating grammar (this ingenious tip comes from Rebecca Johnson, https://twitter.com/johnsonr). For example, "to answer this question, simulations have been performed (BY ZOMBIES)" is clearly passive. Instead, just write "we have simulated". Whenever you spot a passive sentence, try to rephrase it and activate the verb. "X influences Y" is better and shorter than "Y is influenced by X".

- Admittedly, there are times when the passive voice works better. Using the passive voice may be justified when the researcher wants to remove herself or other researchers from the picture. Common examples include "it has been experimentally confirmed that" or "it has been argued that" (especially if you disagree with the argument but do not want to name the culprits).

Also, if you want to stress whatever it is that is being acted upon—that is, the receiver (or victim) of action—use the passive voice. A phrase like "the climate is influenced by greenhouse gases" stresses the word "climate", whereas "greenhouse gases influence the climate" emphasises greenhouse gases.

- Avoid nominalisations—turning verbs into nouns. Nominalisations take the life out of perfectly good verbs. What was once a happy, active verb is shrunk into a sad noun that just sits there, leaving you in need of a replacement verb. These are usually clunkier and duller, like "to carry out", "to perform", "to conduct", or plain "to be". As well as sounding like corporate speak, these verbs make your sentences longer than they need to be. When you happen to come across nominalisations in your draft, rescue and release the original verb from captivity and let it roam free again! Say "we compared" instead of "we performed a comparison", say "we examined" instead of "we conducted an examination", and say "we analysed" instead of "we carried out an analysis".
- To spot a nominalised verb, look for nouns that have a captive verb inside. They often sound like French or Latin and end with "-ion", "-ment", or "-ence". For example, instead of saying "there is a difference between X and Y", you can save 21 characters by writing "X differs from Y".
- Avoid words that end in "-ive". These words tend to be adjectives with a verb inside, struggling to get out. Release the verb! Instead of saying that

"X is indicative of Y", simply say "X indicates that Y."

- Comb your text for clunky expressions that would be simpler and shorter in plain English. For example, "because" is much more effective than "as a consequence of" or "due to the fact that". "Although" works better than "despite the fact that." Don't say "for the purpose of" when you can simply say "to", or "for". "To" is often better than "in order to". For more examples, see http://plainenglish.co.uk/files/alternative.pdf.
- Avoid using jargon and complicated words as a security blanket. Many people seem to think that writing sounds more academic and scientific if it is complicated and full of expressions that no one uses in everyday speech. This is wrong. Science is difficult enough as it is—do not make it any more complicated with your writing. Simple, easy-to-understand writing has authority. It is more difficult to trust a writer who hides her point behind a facade of long sentences and complicated words. Those words feel like smoke and mirrors, tricks to hide the absence of real depth.
- Bonus tip: learn from the masters. Get "*The Elements of Style*" by Strunk & White, and do as the masters tell you. Your readers will thank you for it.

PART IV

IT'S NOT OVER YET

18

HOW TO WRITE THE COVER LETTER

"Brevity is the soul of wit." – William Shakespeare, Hamlet

Now that you have revised and polished your draft, you are the happy owner of a shiny new manuscript. One more step remains before you can submit it to a journal: writing the cover letter.

The cover letter is, in my view, something of a historical remnant, and having to write one is rather annoying. Let me explain.

The purpose of the cover letter is to convince the editor that your manuscript is solid and important and that it fits their journal. But this is something that the abstract should do, especially if it is written in an accessible way. The paper itself should do the job, too. The introduction should contain all the information that the editor needs when deciding whether the paper falls within the journal's scope. The introduction should also reveal whether the results are believable and important enough for your paper to be sent to referees. So why repeat all this in a redundant letter?

This appears to be quite a common view among my

colleagues. But the reality is that most journals still require cover letters, so, unfortunately, you'll have to write one. Here are some thoughts on how best to approach it.

If I was a time-pressed editor, I would greatly appreciate a short, focused cover letter—say three to four paragraphs or a maximum of one page. If such a letter made it clear that the paper's topic fits the scope of my journal, that its results are in some way meaningful, and that the science looks solid enough to warrant more detailed scrutiny by the referees, I would be a happy camper.

Here are two ways to write a short cover letter with only a few paragraphs and less than a page of text.

The better and slightly more adventurous approach follows the inverted-pyramid method that journalists often use for news stories. This may feel difficult at first because most scientists are not used to writing this way. When following the inverted pyramid, you should begin with the most important thing and then proceed towards less important things one by one and in order of decreasing importance. Explain what you have found out in the very first sentence or two before outlining why your finding matters, and only then say something about how you obtained your results. Do not write a detailed explanation of your methods unless they are the key point of the paper: the editor is probably too busy to care, and if not, the details can be found in your manuscript. This way of structuring the letter is particularly suitable for those top-tier journals whose editors desk reject most of the papers that they receive—they do not have the patience to search for the main point if it is buried somewhere on page two of your letter. They want to hear it first and then decide.

As a side note, the inverted-pyramid method should always be used for press releases; those are read by journalists, not scientists, and journalists only get confused if they have to wade through lengthy introductory material before the main point arrives.

The other, more traditional approach is to structure your cover letter in the same way as the abstract or the Introduction. That is, use the hourglass structure: begin with the broad context, and then narrow the scope down and proceed towards your specific research question. After stating the question, describe what you have found out and how and why what you have found out matters. But please be swift and move quickly: the first paragraph for context and question, the second paragraph for the key result, and the third paragraph for significance.

You could also consider writing a hybrid version of the traditional cover letter and the inverted-pyramid story lede. First, state your key result in a single-sentence paragraph. Then follow the structure of the abstract by explaining the context and the question in the second paragraph before adding a more detailed description of the result in the third paragraph and an account of its significance in the fourth.

Whichever method you choose, put emphasis on the implications and impact of your results. The why-does-it-matter part matters more than the how-did-we-do-it part, even if you have used particularly inventive methods. Do not exaggerate the significance of your results—just be honest and explain what your work means. Whenever you feel like typing the word "very", take a deep breath, command your fingers to stop, and jump directly to the next word.

Your cover letter may serve purposes beyond the avoidance of desk rejection. You may also have to state that the manuscript is not under consideration anywhere else or provide similar information. More often than not, there are separate entry fields for such things in journals' online submission systems, so you don't usually have to bother with them when writing the letter.

You may also need to suggest referees, either in the letter or in the online submission system. This is an important choice. The best advice I can give on selecting referees is that you should choose nice people who know about your field—but this is easier said than done. My editor colleagues tell me that the referees suggested by the authors are often no friendlier than the rest—in fact, they're often the opposite. So choose wisely, and if possible ask your most senior co-author for advice.

To summarise: if you have to write a cover letter, keep it simple, keep it short, put important stuff first, and explain why your work matters.

19
HOW TO DEAL WITH REVIEWS

> "...the paper strikes me as trivial and as an addition in the endless series of thrusts beyond unsettled frontiers." — An anonymous referee for *The Strength of Weak Ties* by M. Granovetter, now a classic with over 50,000 citations

Several weeks or months after letting your paper out of your hands and into the cold and hostile world of scientific publishing, the dreaded moment of judgment finally arrives—if you haven't been desk rejected by the editor within a few days, that is. There is a letter from the editor in your inbox, either rejecting your manuscript outright or requesting that you revise it to give it a second chance. This statement is followed by detailed referee comments that may or may not make sense to you (or anyone else, for that matter). In theory, it is also possible to have the paper accepted as it is, but this is very rare: it has happened only once or twice during my career. It is safe to assume that there is more work to do now and that you have to get back to your paper. This has often felt inexplicably painful because the submitted paper has already been neatly wrapped up and archived in my mind. I've gone on to think of other things, only to be told that wrapping has to be torn open—what a pain.

When receiving the letter, my suggestion would be to first take a few deep breaths and to try to calm yourself. This gets easier after dealing with tens of rejections and criticisms, but it never gets easy. That moment of realising that someone is critical of your work is difficult, especially for PhD students to whom a manuscript may represent 100 per cent or 50 per cent of their publication record.

What often happens at this stage is that you rush directly to the referee comments and read through them at top speed. This is because you want to know RIGHT AWAY what it is that the referees think is wrong with your beautiful work. And because you speed through the comments, you only see the surface and only pick up words that criticise your work. Your view is then distorted.

Next, being a human, you get all emotional: angry, embarrassed, frustrated, depressed, or any linear combination of these. You may feel that your work is worthless and therefore that you are worthless. Please don't worry—feeling like this is as normal as it gets. We have all felt like this. This is what a career in science is about. Those who learn to persist and deal with these emotions are the ones who survive. It will get easier, and part of why it gets easier is that you learn that the process of peer-review can be highly random and the referees can be wrong. But often they are not, and you can learn from their criticism and use it to improve your work.

When faced with a letter from the editor where referees criticise your work, my recommendation is to take a few more deep breaths, calm down, and maybe do something else for a while. If the reviewers are very critical, your fight-

or-flight-response prevents you from seeing what is really being said and what the really important problems are, and from assessing how difficult it would be to fix them in a revised version. So, get some distance first. Breathe, get a cup of coffee, look at a video of kittens playing, and only then read the letter again—this time more slowly and analytically.

First, focus on what the editor says because this is the most important thing of all. As an example, my colleagues and I successfully had an article published after the referee hated the paper and asked for a total rewrite from a different and quite alien perspective. This rewrite would have been impossible for us to do, but the editor seemed to like the paper and told us that only a few minor things needed to be done. These two requests were completely at odds with each other, but we chose to listen to the editor instead of the referee (although we picked some of the referee's points and cosmetically revised a few sentences, just in case). The paper was immediately accepted after resubmission.

So, always listen to the editor! If the editor encourages you to submit a revised version, do it—she is already on your side. If the editor uses less encouraging words like "should you wish to resubmit a revised version", do it anyway because even in this case the door is still open.

Sometimes what you receive from the editor is a standard, copy-pasted response in which the editor takes no stance but leaves you to deal with the referees. In this case, just move on to the referees' comments. But if there is anything in the editor's words that you can use, be sure to do so.

Next, read through the referee comments carefully and with an analytical mind. Start by looking at the science and stay positive: did the referees spot any obvious flaws in your work? If so, great, now you can fix those flaws! Did they misunderstand your results? If so, great, this shows that you need to be clearer in your writing (although referees do sometimes miss explanations that are right there in the manuscript). Do your referees require some extra experiments or calculations to back up your conclusion? Do these make sense? If so, great, they will make your paper more solid and you should do them, even if this takes time.

After you have gleaned useful and actionable information from the referee comments, have a look at what is left. There may be comments that you don't understand, comments that you understand but know are wrong, and all kinds of other weird debris. The worst comments often read something like "I am sure that I have seen a similar result somewhere but cannot be bothered to find a reference". But before you react to comments like these, take off your scientist hat for a while and put on your psychologist hat (don't have one? Get one—they are tremendously useful). It pays to remember that referees are human and humans are not analytical machines; they have feelings. Your referees have feelings too. Figure out how they feel and what to do about it.

Try to see the world through the eyes of the referees. Read the negative comments again, and try to understand what the referees think, feel, and expect from you. Do they feel irritated? Left out? Bored? Insecure and needing to bolster their confidence? Confused? Just plain cranky?

The key to getting your revised paper accepted is to understand what makes the referees happy and then to give them this (while maintaining your scientific integrity, of course).

One way of making the referees happy is to always

acknowledge that they have been heard, whatever it is that they say. Never treat a referee with disrespect, even when what they propose is wrong or when they have misunderstood things because they obviously didn't bother to read your manuscript carefully enough. Always give them something.

If the referee misunderstood something that is obvious, write another sentence about the obvious issue and insert it somewhere. Thank the referee for pointing out that your paper was not clear enough on the issue, and tell her that you have now added a clarifying sentence. If the referee's comment or question is so confused that you can't even figure out what it is that the referee wants from you, do something about it anyway. Pick some sentence or paragraph that might be related to whatever is confusing the referee. Then rewrite it: try to make it clearer, or at least rearrange the words. Then answer the referee politely: tell them that, to the best of your understanding, her problem was probably related to the issue discussed in this sentence or paragraph and that you have now tried to make it clear. At times, the referee will be confused enough not to know herself what the original issue was, and will gladly accept this act of repentance from you.

The above does not mean that you should always do what the referee tells you to do. If the referee asks you to do something that you feel is wrong or doesn't make sense, don't do it. You should, however, present a detailed counter-argument in the rebuttal letter. But, if at all possible, change something in the text—however small—so that the referee will feel that she has been heard.

Referees commonly come up with speculative ideas. While this can be genuinely helpful, it can also be a nuisance if those ideas are tangential to what you are doing.

If the referee asks you to do something that sounds sensible but is clearly outside the scope of your paper, thank the referee and insert a sentence into your Discussion section where you mention that it would be useful to do that in the future.

It may be that the referee is both hostile and wrong: you know that your results hold, but the referee does not believe you and does not want to believe you. In this case, you must take a stand and defend your results. Be constructive and respectful. Why does the referee act this way? If it is because the referee wants more evidence, apologise for the lack of evidence and provide more (even if there was enough already). If it is because the referee feels left out (often indicated by requests to cite some of her own work), you can give her credit for some earlier work in the introduction or discussion, but you should not cite papers only because you are forced to do so.

There may even be more nefarious reasons for referee hostility, such as the referee trying to block competition. That may be hard to detect or disentangle from general grumpiness (perhaps the referee just had low blood glucose levels, you never know), but if you do suspect something like a blocking attempt, be polite but firm in your responses—remember that they are also meant for the editor's eyes. If someone has decided to block your work then you won't be able to turn that person's head, but the editor might be able to spot what is happening. And if a referee is clearly being unethical, you should confidentially let the editor know about this.

However, more often than not, your referees are not nasty for the sake of it. If you treat them with respect, make sure that they feel they have been heard, and if you always give them something, then they will be happy and they will accept your paper more willingly. And if not, there are always other journals.

In a nutshell, when you receive the dreaded letter from the editor with referee comments, do the following:

1) Breathe. Relax. Get a cup of coffee.

2) Read the editor's comments. Figure out if there is anything useful in there and do whatever the editor wants from you.

3) Read the referee comments slowly and analytically, first focusing on those that are useful: fixing flaws, carrying out extra experiments, backing up your claims with more conclusive evidence, being clearer.

4) Read the rest of the referee comments with your psychologist hat on. What would make the referees happy?

5) Fix flaws, carry out new experiments. Revise your paper accordingly. For every referee comment, change something in the manuscript.

6) Write a rebuttal letter, thanking the editor and the referees for their comments. Respond politely and in detail to every comment, even the silly ones; if the referees are confused, act as if this was your fault. Tell them what you have changed in response to their comment (this can be just a single sentence).

7) Make sure that the editor's concerns are explicitly addressed in your rebuttal letter.

8) Make sure that each and every referee comment is addressed in detail, including requests that you chose not to fulfil. Unlike the cover letter, the rebuttal letter does not need to be short. The longer and more detailed, the better.

9) Resubmit the paper and the rebuttal letter. Keep your fingers crossed.

SUMMARY

For quick reference, here are the steps of the method introduced in this book.

I: Develop the story of the paper

1. Choose your key point and main conclusion. Pick the single most important result that you have. Condense it into one or two sentences.

2. Choose what else goes in. Keep only those results that support your key result or illustrate its significance. Categorise the results according to their role. Leave the rest out or tuck them in the Supplementary Information document.

3. Write the abstract first. Follow the hourglass format: broad context, narrower context, exact research question, your solution, implications of your result on your (sub)field, implications of your result in a broader context.

4. Draft the title of the manuscript. Short titles work better; make sure that the title conveys the point of your paper and aligns seamlessly with your abstract.

II: Outline the paper

5. Sketch the outline of your paper. Choose the order of presentation for your results. Choose your figures and write

first versions of figure captions. For each section (Introduction, Methods, Results, Discussion), write down the main points and plan their order of appearance.

6. Sketch the paragraphs of your paper. Follow your outline and transform it into paragraph-sized blocks, so that every paragraph makes one point only. Write down the point of each paragraph, together with bullet points for the paragraph's contents; you may already want to sketch the key sentences of the paragraph too. List references if they already come to mind.

III: Write and revise your draft

7. Quickly complete your paragraphs and write a first rough draft. Stick to the point with each paragraph. Make sure that the first sentence of the paragraph explains what the paragraph is about. Make sure that each sentence leads to the next. Make sure that each paragraph leads to the next.

8. Revise your first draft for structure and content. Make sure that the paper is focused and readable, that there is enough context and background, and that the story is about an important question that your results answer.

9. Revise your first draft focusing on sentences and words. Make sure that each sentence is logical and that its meaning becomes apparent early on. Keep the subject and verb close. Use the stress positions at the ends of sentences. Write in the active voice. Split long sentences. Remove superfluous words. Avoid turning verbs into nouns.

10. Do several rounds of revisions. Focus on clarity, readability, and a good story. Edit jointly with co-authors. Ask for comments from your supervisor and your colleagues. Use test readers for identifying hard-to-spot problems.

IV: Submit and deal with reviews

11. Write a cover letter if you have to. Make it short and to-the-point.

12. Submit your paper and wait for referee comments.

13. When the referee comments arrive, take a deep breath; if you get an emotional reaction, wait until you are calmer. Try to gauge what the editor wants from you and how to make the referees happy. Revise your paper. Write a long, detailed rebuttal letter and resubmit.

AFTERWORD

"The pen is mightier than the sword if the sword is very short, and the pen is very sharp." — Terry Pratchett

"If you don't have time to read, you don't have the time (or the tools) to write. Simple as that." — Stephen King

Thank you for reading this book; I hope it was helpful.

Whatever stage you are at in your career as a scientist, writing is a craft that is worth putting some effort into. It is also a craft that anyone can learn. While some seem to have a head start through natural talent, each and every one of us can learn to write. You just need to decide to learn, and then to put in the hours.

At the beginning of my scientific career, I sort of enjoyed writing but found it very difficult: there seemed to be all these unstated rules and fancy words that you had to know and use to make your writing sound "academic". There were no books or learning materials, and there was no online guidance. I remember taking a course in "Technical

English" that in the end was focused on writing technical manuals. That was pretty much it. Because of this, I had to learn by imitation, by reading papers and attempting to mimic their style. I learned things I had to unlearn later—I wouldn't dare look at my earliest papers now. I certainly didn't live as I teach.

The first book on scientific writing that I came across a few years back was Joshua Schimel's *Writing Science*. Reading the book was a revelation. I was already a professor in charge of supervising many graduate students, who all struggled with writing, just as all of us scientists do. I found it hard to help my students to improve their writing because I didn't have a system myself—I just winged it, so to speak. Schimel's book opened my eyes: there are formulas, systems, and rules that make writing easier. And papers can be stories!

If, after reading this book, you want to continue your journey towards writing focused, readable, and brilliant papers, my suggestion is to have a look at Schimel's book. I would also recommend *The Elements of Style* by Strunk & White and *On Writing Well* by William Zinsser. You should also read a lot of scientific papers—critically and analytically. How do great papers structure their narratives? Why are some papers easy to follow? Why are some papers impossible to grasp (hint: it's not you, it's them)? Read and study papers, and analyse what works, what doesn't, and why.

Then do a lot of good science, sharpen your pen (or configure your Emacs), write, rewrite, and rewrite some more.

Jari Saramäki, Espoo, Finland, October 2018

P.S. Can I ask for a favour? I'd really appreciate a short review on Amazon—reviews help new readers find this book!